THE 90-DAY AI ADVANTAGE

How the Small and Mid-Market Owner Builds an AI-Powered Business — Without Losing What Makes It Theirs

LEE ALLEN MILLER

Bestselling author of Entrepreneurship God's Way and The Faith Media Revolution

The 90-Day AI Advantage

ISDN: 979-8-9921074-3-2 Paperback

ISDN: 979-8-9921074-3-2 Kindle

Published by the book division of

MSGPR Ltd Co · Lufkin, Texas

Case studies are drawn from real engagements; names and figures have been changed to protect the privacy of the businesses involved.

CONTENTS

CHAPTER ONE

Why You Can't Wait

A Book for the Owner in the Middle

Let me tell you who this book is for, because it is not for everyone, and saying so plainly at the start will save us both time.

It is not written for the giant corporation with an innovation department and a budget line for experiments. That company does not need this book; it has people whose entire job is to think about what comes next. And it is not written for the technology enthusiast who already spends weekends trying every new tool that appears — that reader is having a fine time and needs no help from me.

This book is written for the owner in the middle. The person who runs a real business — a good one, built over years, employing real people, serving real customers — a business large enough to matter and small enough that there is no department to hand the future to. If you are that owner, then you are the person I have been thinking about on every page, and I want to begin by being honest with you about why this book exists and why its title contains the words it does.

— ❖ —

More to Do, Fewer to Do It

Start with something you already know in your bones, because you live it. The business you run today is being asked to do more than the same business was asked to do a decade ago — and it is very likely being asked to do it with fewer people, or at least with no more.

The pattern is everywhere, across nearly every industry. The work has multiplied. There is more to track, more to comply with, more to market, more to answer, more channels to be present on, more that customers expect as a matter of course. And the team has not multiplied to match. In many businesses it has shrunk. The owner and the people around them are doing the work of a larger organization with the headcount of a smaller one, and they are holding it together by sheer effort and long hours.

> *The work multiplied. The team did not. Most owners are holding a larger organization together with a smaller one's headcount — by sheer effort and long hours.*

Most owners respond to this the only way effort alone allows. They triage. They handle whatever is on fire today, push everything else to next quarter, and next quarter never comes — it arrives already full of its own fires. The important-but-not-urgent work, the building and improving and getting-ahead work, slides perpetually down the list. It is not that owners do not see what would make the business better. It is that there is no capacity left to do it. Every hand is already full.

That is the squeeze, and if you are reading this book it is almost certainly the water you swim in. Hold it in mind, because it is the precise problem the rest of these pages are built to solve.

— ❖ —

What AI Actually Changes

Now set the squeeze beside what artificial intelligence has genuinely become — and let me be careful here, because almost everything written about AI for business owners errs in one of two directions, and both errors will cost you.

The first error is breathless overstatement — the promise that AI will transform everything overnight, run your business for you, change the world by Tuesday. That is hype, and you are right to distrust it. The second error is the quiet dismissal — the conclusion, often reached after one disappointing experiment, that AI is overrated, a passing fashion, not really built for a business like yours. That, too, is wrong, and it is the more expensive of the two mistakes.

Here is the honest middle, the ground this whole book stands on. AI is not magic and it is not a fad. It is a genuine, practical tool that does one thing of enormous value to a business under the squeeze: it takes on work. Real work. The repetitive, time-consuming, mechanical work that currently consumes your best people — the drafting, the documentation, the data-handling, the routine production — the very tasks that have been burying the important-but-not-urgent under the urgent. AI is not, for the businesses in this book, a way to replace people. It is a way to give a small team the operational capacity of a larger one.

> *AI is not magic and it is not a fad. It is a practical way to give a small team the operational capacity of a larger one — which is exactly what the squeeze requires.*

Read that against the squeeze and the fit is exact. The squeeze is too much work and not enough people. AI is a tool that does work and does not need to be hired. It is not an exaggeration to say that AI may be the first thing to come along in a generation that genuinely answers the central problem of the small and mid-market business. That is a strong claim. The rest of this book is the case for it.

Why the Title Says "Can't Wait"

If AI is genuinely useful, a reasonable owner might still ask the most reasonable question there is: why now? Why not let the technology settle, let others work out the wrinkles, and adopt it in a year or two when it is more proven and more obvious how to proceed? That instinct — to let something mature before committing — has served careful owners well for a long time. I want to explain, honestly, why I believe it does not serve you here.

It is not because the tools will stop improving if you wait; they will keep improving whether you adopt them or not. It is not about fear, and you will find no doom in this book. It is about one specific, unsentimental fact: your competitors face the same squeeze you do, and AI is equally available to them. The advantage in this technology does not go to the business with the most resources. It goes to the business that moves first — that begins building capacity, and learning, and pulling ahead, while others are still waiting for a more convenient moment.

Picture two businesses, much alike, both under the squeeze. One begins — carefully, modestly, but begins. Over a year it recovers real capacity, frees its people from a measure of the grind, and quietly grows more capable month by month. The other waits for the perfect time. A year on, the gap between them is not small, and it is not closing. It is widening, and it will go on widening, because capacity compounds. The first business is not winning because it had more money. It is winning because it started, and the other did not.

That is the whole meaning of the words *can't wait* on the cover of this book. Not panic. Not pressure. Simply the plain truth that there is a real and compounding cost to standing still — and that the cost is paid quietly, in advantages handed to the competitor who chose to begin. Waiting feels safe. In this particular case, waiting is the risk.

What This Book Will and Will Not Do

Let me be equally plain about what you are holding, so your expectations are set correctly from the first chapter.

This book will not hand you a list of tools to buy. It cannot honestly do that — the specific tools change too quickly for any printed page to keep up, and a book built on this month's products would be wrong within the year. What this book offers instead is more durable and more valuable: a way of thinking, and a clear path to walk. It will give you a framework — ninety days, three phases — that any owner can follow regardless of which tools happen to be best when you read this. It will show you how to make AI genuinely understand your particular business. It will answer, honestly, the objections you are already forming. And it will do all of this in the plainest language I can manage, because you do not have time for jargon and you should not have to learn a new vocabulary to run your own company well.

It is also a short book on purpose. You are busy — that is the entire premise of these pages — and a long book would be a strange way to honor that. Every chapter is meant to be read and then used. This is a working document, not a treatise.

And one more thing you should know about the ground this book stands on. I am a business owner, and I am also a person of faith, and I do not believe those two facts sit in separate rooms. I believe a business is something built and stewarded — that the people in it are owed our care, that the thing we build is meant to last and to be handed forward, and that new tools are to be taken up wisely rather than chased or feared. That conviction runs quietly through this whole book. You will not find it preached at you. But it is why the framework here refuses to put your people second to efficiency, why it treats the recovered time as something to be *stewarded* rather than merely spent, and why it cares as much about the business you hand to whoever comes after you as about the quarter in front of you.

The Promise of the Next Eight Chapters

So here is what the rest of this book will do, and the promise it makes to you.

The next chapter looks honestly at why capable owners — owners exactly like you — so often freeze on this decision, and why the reason has almost nothing to do with the technology. After that, the book lays out its foundation: the idea that makes AI genuinely useful to your specific business rather than generically capable. Then comes the heart of it — the 90-day framework, three phases of thirty days each, the practical path from where you stand today to a measurably stronger business one quarter from now. You will see the framework proven in real companies. You will get straight answers to every objection worth raising. And you will finish with a first step small enough to take this week.

Nothing in these pages will ask you to be a technologist, to spend heavily, or to risk what you have built. The entire approach is designed for the careful owner — one small, testable step at a time, each result earning the next. That is the promise: not a transformation you must gamble on, but a path you can walk, beginning now, at a pace you control.

> *Not a transformation you must gamble on. A path you can walk — beginning now, one testable step at a time, at a pace you control.*

The squeeze is real, and it is not going to ease on its own. But for the first time in a long while, there is a genuine answer to it — one that is available to you, today, regardless of your size or your budget. The only question this book cannot answer for you is whether you will begin. Everything else, it will walk you through. Turn the page, and let us look honestly at what has been holding you back.

CHAPTER TWO

The Real Problem Isn't the Tools

Why Capable Owners Freeze

If the problem were really the tools, this would be a much easier book to write – and you would not need it. The AI tools available to any business owner today are remarkable, inexpensive, and getting better every month. They are not the obstacle. If capability were all that stood between you and an AI-powered business, you would already have one.

So if the tools are not the problem, something else is. And in my experience – across many businesses and many owners – the real obstacle is almost never technical. It is the quiet fact that the owner is facing the decision alone. To understand why AI adoption stalls in capable, serious businesses, you have to understand something that has very little to do with AI at all. You have to understand the particular isolation of the person at the top.

Nobody in the Room

There is a condition that comes with running a business, and almost no one names it out loud, because naming it feels like weakness. It is this: the owner makes the largest, most consequential decisions of the entire enterprise with, very often, nobody in the room.

Think about who surrounds a business owner, and notice that every one of them has a stake. Your employees depend on you for their living — they cannot be fully neutral, and they should not be expected to be. Your family is affected by every major call you make. Vendors want your business. Consultants want to be hired again. Even well-meaning friends in the same industry are, somewhere in the background, competitors. Each of these people may care about you genuinely. Not one of them can be the thing you actually need at the moment of a hard decision: a peer who understands the weight of it, has no stake in the outcome, and owes you nothing but honesty.

> *The owner makes the largest decisions of the whole enterprise with, very often, nobody in the room. That isolation is the real problem — and AI is simply the latest decision it touches.*

This is the isolation of leadership, and it is not a personal failing. It is structural. It comes with the chair. The higher you sit in an organization, the fewer people exist who can speak to you as a true equal about the things that matter most — and the more those decisions cost when they go wrong. Most owners simply absorb this. They carry the weight, they make the calls alone, and they tell themselves that solitary decision-making is just what leadership is.

And here is why a book about AI must begin here. The decision about whether and how to bring AI into your business is exactly this kind of decision — large, consequential, unfamiliar, and faced alone. It is not, at its root, a technology problem. It is one more heavy call landing on the desk of someone who has no neutral peer to think it through with. That is the real reason capable owners freeze. Not because AI is too complicated. Because they are carrying the question by themselves, and a question carried alone tends not to move.

— ❖ —

The Three Ways the Freeze Looks

Isolation around the AI decision does not announce itself. It disguises itself as something more respectable, and it usually takes one of three forms. You may recognize your own business in one of them.

The first is endless research. The owner reads, and reads, and reads. More articles, more videos, another webinar, another conversation. It feels like progress because it feels like diligence — but months pass and nothing in the business has actually changed. This is not genuine research. It is a holding pattern. When a person has no trusted peer to help them decide, they substitute the gathering of information for the making of a decision, and they can gather forever.

The second is the opposite — the impulsive lurch. Tired of the paralysis, the owner overcorrects. They buy six tools in a week, announce a sweeping new direction, and act decisively — and the chapter on the Foundation phase will show you exactly how that ends. The lurch is isolation's other face. With no one to pressure-test the move, the owner swings from doing nothing to doing too much, and neither extreme is a decision well made.

The third is the quietest, and the most common: indefinite avoidance. The owner simply does not engage. AI goes onto the someday list, filed next to all the other important things there is no bandwidth to think about properly. It is not a decision to ignore AI. It is the absence of a decision — and the absence of a decision is, in a changing market, a decision all the same. It just is not one anybody chose.

Look honestly at those three, and notice what they share. Not one of them is a technology problem. Endless research, the impulsive lurch, indefinite avoidance — every one of them is what a capable person does when a heavy decision has no one beside it. The freeze is not about AI. It is about being alone with the question.

— ❖ —

An Exercise for the Empty Room

If the problem is that you face the decision alone, then the beginning of a solution is to find a way to bring other perspectives into the room — even when, in plain fact, you are the only one sitting there. There is an exercise I give owners for exactly this, and it is worth doing with a real pen and real paper before you read another chapter.

Take the decision in front of you — for most readers of this book, the question of whether and how to begin with AI — and write about it three separate times, each time from a different chair.

Chair One. The Operator.
Write what the numbers say. Pure arithmetic, no feeling. What is the financial case for acting? What does inaction cost? What does the math, and only the math, tell you to do? Set your gut aside entirely and let the spreadsheet speak.

Chair Two. The Human.
Now write what your gut says. How does this decision touch the people involved — your team, your family, your customers? What feels right, and what feels wrong, and what are you afraid of? Set the

spreadsheet aside entirely. This chair is only the people.

Chair Three. The Outsider.
Finally, write what a person with no stake in the outcome would say. Someone who does not work for you, is not married to you, does not need your approval or your business. Someone who can see the situation plainly because they are not standing inside it. What would that person tell you that you have not been able to tell yourself?

Now read all three back to back. Two things tend to happen, and both are useful.

First, the Operator and the Human will often disagree — the numbers point one way, the gut another. Do not be troubled by that. That very tension is usually why the decision has felt stuck. You were not failing to decide; you were holding two true things that pulled in different directions, with no way to set them side by side.

Second — and this is the heart of the exercise — Chair Three is where the clarity hides. When you force yourself to write from outside your own situation, you will almost always find yourself naming the thing you have quietly known all along and have not let yourself say. Because here is the truth about owners and hard decisions, and it is worth stating plainly: most owners who are stuck are not stuck because they do not know what to do. They are stuck because they know exactly what to do, and they have not had anyone give them permission to do it.

> *Most owners who are stuck are not stuck because they don't know what to do. They are stuck because they know — and no one has given them permission to do it.*

I will be honest with you about what this exercise is and is not. It is not a true substitute for a real person in Chair Three — a genuine, trusted, agenda-free peer is a different and greater thing, and the last chapter of this book will have more to say about that. But the exercise will do something real now. It will show you what has been stuck, and why. And for many owners, the act of writing the Outsider's paragraph is enough to break the freeze — to turn months of motionless research or avoidance into a decision, and a decision into a first step.

— ❖ —

What This Means for the Rest of This Book

If you have followed the argument of this chapter, then the rest of the book should look a little different than it did a few pages ago.

Everything that follows — the 90-day framework, the Blueprint, the case studies, the honest answers to your objections — was designed by someone who understands that the real obstacle was never the technology. It was designed for an owner who is intelligent, capable, and carrying the decision alone. That is why the framework refuses to ask for one large, irreversible bet. A person facing a frightening decision in an empty room cannot safely make a leap. But that same person can take one small, testable step, see the result, and take another. The framework is built the way it is precisely because of the isolation this chapter has described. It is a path designed to be walked by someone who has to begin it by themselves.

And there is a quiet word of grace in that, worth saying directly. If you have felt stuck on this — if you have done the endless research, or the avoidance, or watched yourself swing between them — the explanation is not that you lack the ability. It is that you have been doing a genuinely hard thing without the one resource that makes hard things manageable: another mind beside yours. That is not weakness. It is simply the cost of the chair you sit in. And it is a cost this book is built to help you carry.

So the question, as the next chapter begins to lay out the framework itself, is no longer whether AI is too complicated for your business. It is not. The question is whether you will stop carrying the decision alone — and take the first small step. The next chapter shows you the foundation everything else is built on. The one after it shows you where to put your feet.

CHAPTER THREE

The AI DNA Blueprint™

Teaching AI to Know Your Business

Sit down at an AI tool for the first time, type an honest request — write a follow-up note to a customer, draft a description of what we do, summarize this for my team — and you will get something back almost instantly. It will be grammatical. It will be competent. And it will be, in a way that is hard to name at first, wrong.

Not wrong in its facts. Wrong in the way a stranger is wrong when they try to speak for you. It does not sound like your business. It does not reference what you actually do, the way you actually describe it. It reaches for the generic phrasing that could belong to any company in your industry, because that is exactly what it knows — your industry in general, and your business not at all. Most owners meet AI, have this exact experience, and quietly conclude the tool is overrated.

The tool is not overrated. It is uninformed. And the difference between those two things is the most important idea in this book — because one of them is a dead end and the other is a problem with a name and a solution. The solution is what this chapter is about. I call it the AI DNA Blueprint.

— ❖ —

The Most Capable New Hire You Have Ever Made

Here is the most useful way I know to think about an AI tool, and it reframes the whole problem.

Imagine you have just hired someone genuinely brilliant. They are quick, tireless, widely read, and able to turn their hand to almost anything you put in front of them — writing, analysis, drafting, organizing. By raw capability, the best hire you have ever made. But it is their first morning. They have never seen your business. They do not know your customers, your services, your history, the way you speak, the standards you hold, the things you would never say. They are immensely capable and completely uninformed, both at the same time.

What would you expect from that person on day one? Exactly what you got from the AI tool. Competent, generic work that does not yet sound like your company — because how could it? You have not told them anything. No one would hire a brilliant employee, skip every word of onboarding, hand them a task in the first ten minutes, and then judge the company on the result. Yet that is precisely how almost every business owner first uses AI: they hire the brilliant stranger and then expect them to already know everything.

> *An AI tool is not overrated. It is a brilliant new hire on their first morning — immensely capable, and completely uninformed. The fix is not a better hire. It is onboarding.*

Once you see it this way, the path forward is obvious. The fix for a brilliant but uninformed new employee was never to go looking for a different employee. The fix is to onboard the one you have — to teach them your business until they understand it. The AI DNA Blueprint is simply that onboarding, written down. It is the document that turns a brilliant stranger into something far more valuable: an insider.

— ❖ —

What the Blueprint Is

The AI DNA Blueprint is a single, structured document that teaches an AI tool everything it needs to know about your business in order to work as though it belonged there.

The name is chosen with care. It is a blueprint because it is structured and architectural — not a pile of notes but an organized plan, the foundation that everything else is built on. And it is DNA because it captures what is genuinely particular to your company. A template cannot produce it. It cannot be copied from a competitor. It is the genetic code of this business and no other — which is exactly why, once an AI tool has it, the tool stops sounding like everyone and starts sounding like you.

In practice it is not exotic. It is a written record of the things that, in a healthy business, usually live only in the owner's head and a few long-tenured employees' instincts. Several kinds of knowledge belong in it:

The Knowledge a Blueprint Captures

1. **Your voice.** How your business actually sounds — warm or precise, plain-spoken or formal, the words you favor and the tone you would never use. This is what makes AI output read as yours rather than as a stranger's.
2. **Your services.** What you actually offer, described the way you describe it — not the generic category an outsider would assume, but your real work in your real words.
3. **Your customers.** Who you serve, what they care about, what they worry about, how they make decisions. An AI tool that understands your customer writes for that person instead of for a vague average.
4. **Your operations.** How the work actually moves through your business — the workflows,

the steps, the way things get done. This is what lets AI assist with the work rather than just talk about it.

5. **Your standards and your terminology.** The language particular to your business, and the lines you hold — what good work looks like to you, and what you would never put your name on.

None of this is mysterious. Every one of these things is already true about your business. The Blueprint does not invent them. It simply gathers what is currently scattered — across your memory, your team's habits, your website, your files — into one place an AI tool can actually be given.

Why the Blueprint Changes Everything Downstream

It would be easy to hear all this and think the Blueprint is a nice refinement — a way to make AI output a little better. That undersells it considerably. The Blueprint is not a refinement. It is the foundation, and there are three reasons it deserves that word.

The first is quality. The single largest gap between disappointing AI results and genuinely useful ones is not the tool and not the skill of the person typing. It is context. An AI tool given real knowledge of your business produces work that is better in kind, not just in degree — work an experienced member of your team would recognize as belonging to the company. Without that knowledge, you are forever correcting a stranger. With it, you are directing an insider.

The second is leverage, and this is the one owners feel most. Without a Blueprint, every AI session starts from nothing. You re-explain your business, re-establish your voice, re-supply the context, every single time — and that re-explaining quietly eats much of the time AI was supposed to save. A Blueprint is built once and then used everywhere. The effort is paid a single time and the benefit is collected on every task afterward. That is the definition of leverage, and it is why the Blueprint multiplies the value of everything else in this book.

> *Without a Blueprint, every AI session starts from nothing — you re-explain your business every time. The Blueprint is built once and used everywhere. That is leverage.*

The third reason reaches further than efficiency. A Blueprint is, quietly, an act of preservation. The knowledge it captures — your voice, your standards, the hard-won understanding of how your business actually works — is, in most companies, dangerously informal. It lives in the owner's head and in the instincts of a few people who have been there longest. It has never been written down, which means it can be lost: when someone retires, when someone leaves, when memory simply fades. The act of building a Blueprint forces that knowledge into words. And knowledge in words is knowledge the business owns — something that can be taught to a new employee, handed forward, kept. For an owner who thinks in terms of legacy, of building something meant to last beyond any one person, that is not a side benefit. It may be the most important thing the Blueprint does.

Where the Blueprint Sits in the 90 Days

Chapter One promised you the 90-day framework — three phases: Foundation, Integration, Expansion — and the next three chapters walk it in full. It is fair to ask, before we get there, how the Blueprint relates to it. The answer is simple: the Blueprint is what makes the framework work as well as it can.

Consider the Foundation phase, which the next chapter lays out in full. You will choose one costly workflow and put AI to work on it. You can do that with no Blueprint at all — and it will help, because even an uninformed AI tool saves time on the right task. But an AI tool working on that same workflow with a Blueprint behind it is working as an insider rather than a stranger, and the result is better, faster, and needs less correction. The Blueprint does not compete with the framework. It is the difference between running the framework and running it well.

This is why a Blueprint, even in a simple early form, belongs near the start of the journey. It does not need to be complete before you begin – it can start small, covering only the part of the business your first workflow touches, and grow from there. But the principle holds throughout: the more your AI tools genuinely know your business, the more every phase of the framework returns. Foundation, Integration, Expansion – each of them runs on context, and the Blueprint is how context gets supplied.

— ❖ —

An Honest Word About Building One

You may be wondering how hard a Blueprint is to build, and you deserve a straight answer.

The concept is not complicated – this chapter has given it to you in full, and nothing about it is hidden. A capable owner who understands what a Blueprint is, and why it matters, can absolutely begin assembling one. Start by writing down, plainly, the five kinds of knowledge described earlier as they apply to your business. That alone – your voice, your services, your customers, your operations, your standards, set down in one document – will already change the quality of what your AI tools give back. I would encourage any reader to do exactly that, and soon.

What is genuinely harder is not the concept but the craft of it. Building a Blueprint that is truly thorough — that captures the voice precisely, that maps the operations faithfully, that anticipates the workflows AI will be asked to support — takes a particular kind of attention, and it is the kind of work that benefits from a practiced outside eye. An owner is often too close to their own business to see what an outsider would need to be told. That is honest, and it is why a Blueprint is something some owners choose to build with guided help and others choose to build themselves over time. Either path is legitimate. What is not legitimate is skipping it — hiring the brilliant stranger and never onboarding them at all.

So take from this chapter the thing that matters most, which is not a service but an understanding. Your AI tools are not underperforming because they are weak. They are underperforming because they do not yet know your business. Give them that knowledge — in whatever form and at whatever pace suits you — and everything else in this book works better. Withhold it, and you will spend years correcting a stranger and calling it the tool's fault.

With the Blueprint understood, you have the foundation the whole framework was waiting for. The next chapter returns to the 90 days themselves and begins where every real implementation begins — the first thirty days, and the discipline of choosing just one thing.

CHAPTER FOUR

Days 1–30
Foundation

Pick One Thing

The most important sentence in this entire book is also the shortest. Here it is: pick one thing.

Not five things. Not a comprehensive AI strategy. Not a transformation initiative with a steering committee and a slide deck. One thing. The first thirty days of this framework have exactly one job, and that job is to take a single, specific, painful workflow in your business and put AI to work on it — well enough that by the end of the month you can point at a result and say, plainly, that worked.

That sounds modest. It is supposed to. The modesty is the strategy. Because almost everything that goes wrong with AI in a small or mid-market business goes wrong for the opposite reason — too much, too fast, too broad. Before we build the Foundation phase, you need to understand the failure it is designed to prevent.

— ❖ —

Why Most AI Efforts Quietly Die

I have watched a great many businesses begin with AI, and the ones that fail almost never fail for the reasons they feared. They do not fail because the technology was too hard, or because their people were not smart enough, or because they picked the wrong product. They fail because they tried to do everything at once.

The pattern is so consistent it is almost a script. An owner gets excited — rightly excited — reads a few articles, talks to a few people, and decides this is the year the business gets serious about AI. So they buy six tools. They sign up for subscriptions across marketing, operations, customer service, and analytics. They send an email to the team announcing the new direction. And then they go back to running the business, because the business still needs running.

Three months later, every one of those six subscriptions is gathering dust. Nobody was ever really trained on any of them. No single workflow was ever truly changed. The team, watching all of this, draws the quiet conclusion that AI was a phase — and the next time the owner brings it up, the room is a little more skeptical than before. The owner did not get a failed experiment. They got a failed experiment plus a more cynical team. That is worse than not starting.

> *Failed AI efforts almost never die from bad technology. They die from too much at once — six tools bought, nobody trained, every subscription gathering dust by spring.*

The lesson is not that those owners lacked discipline. Most of them are disciplined people; that is how they built a business in the first place. The lesson is that breadth feels like seriousness and is actually the enemy of it. Buying six tools feels decisive. It feels like real commitment. But commitment spread across six fronts is not commitment — it is dilution wearing commitment's clothing. Real seriousness, in the first thirty days, looks almost disappointingly small.

The Single Pain Point Principle

So here is the rule that governs the entire Foundation phase. I call it the Single Pain Point Principle, and it is simple enough to say in one breath: in the first thirty days, you change exactly one workflow, and you do not move on until that one is genuinely working.

One workflow. Not one department, not one category of work — one specific, nameable workflow that someone in your business does on a regular, repeating basis and quietly dreads. The narrowness is not a limitation you are tolerating. It is the mechanism. A single workflow is small enough that you can actually pay attention to it, small enough that you can train someone properly, small enough that you can see clearly whether it worked, and small enough that the cost of being wrong is a month rather than a year.

And a single workflow done well does something that six tools done poorly never can. It produces proof. At the end of thirty days you are not holding a theory about what AI might do for your business. You are holding a result — a real one, in your own operation, with your own people. That proof is the thing every later phase is built on, and it is the thing that turns a skeptical team into a willing one. You cannot argue a team into belief. You can only show them.

> *In the first thirty days you change exactly one workflow — and you do not move on until it genuinely works. The narrowness is not the limitation. It is the mechanism.*

There is a stewardship principle underneath this, and it is worth saying directly for those who think the way I do. Scripture has a great deal to say about the person who is faithful with a little being trusted with much, and almost nothing good to say about the person who grasps at much and is faithful with none. The Single Pain Point Principle is simply that wisdom applied to technology. Prove yourself faithful with one workflow. Then, and only then, take up the next.

— ❖ —

Finding the Right One: Ask the People Who Feel It

If the first thirty days hinge on choosing one workflow, then the choosing matters. And here the most common mistake is to make the decision from behind the owner's desk.

An owner has a view of the business, but it is a view from a particular altitude. You see the shape of things, the numbers, the strategy. What you often cannot see clearly — because you are not the one doing it — is exactly which repetitive task is quietly bleeding hours out of someone's week. The person who knows that is the person doing the task. So the choosing has to start with them.

Sit down with your actual operators. Not just your managers — the people whose hands are on the work. And ask them one direct question: what is the single biggest time thief in your day? Then be quiet and let them answer. You may be surprised. The task that is costing the most is frequently not the one that looks most important from above. It is usually something repetitive, formatted, and unglamorous — the kind of work that never makes it onto a strategy document precisely because it is so routine, and so is never examined.

For most small and mid-market businesses, the honest answers tend to cluster in a few familiar places. There is *the documentation and paperwork load* — reports, records, summaries, compliance forms, the routine writing that has to happen and that nobody enjoys. There is *the communication load* — the drafting of similar emails over and over, responding to the same categories of customer questions, producing the regular updates a business has to send. There is *the data and reconciliation load* — pulling numbers from several systems and lining them up, the monthly close, the matching of one record against another. And there is *the content load* — the marketing posts, the newsletters, the proposals, the recurring material a business has to keep producing to stay visible.

Your business has its own version of these. The point is not the list. The point is that you find the one task, in your actual operation, that is both genuinely repetitive and genuinely costly — and that you find it by asking the person who lives with it, not by guessing from the top.

Three Tests for the Right First Workflow

Is it repetitive? The best first workflow is one that happens again and again on a predictable rhythm — weekly, or monthly, or every time a certain event occurs. Repetition is what makes a result trustworthy. A task that happens once is hard to learn from; a task that happens every week teaches you something every week.

Is it costly? It should consume real, measurable time — ideally several hours a week of a skilled person's attention. You want the savings to be large enough that nobody can wave them away. Saving twenty minutes proves little. Saving most of a day proves a great deal.

Is it low-risk? The first workflow should not be the one where a mistake is catastrophic. Choose something internal and recoverable for your first attempt — a draft a person still reviews, a report someone still checks — not the single most sensitive thing your business does. There is time for the harder cases. The first one is for learning.

A workflow that passes all three tests — repetitive, costly, and low-risk — is your Foundation. Pick it, write it down, and commit to it for thirty days.

The First Thirty Days, Step by Step

With the workflow chosen, the month itself has a simple shape. You do not need a project plan. You need four weeks and a willingness to pay attention.

Week one is for the conversation and the choice. Have the discussion with your operators. Land on the one workflow. Then — and this matters — resist the urge to research ten tools. Spend a modest, bounded amount of time identifying one tool suited to that one workflow. The point of week one is not a perfect tool. It is a chosen tool and a chosen task.

Week two is for the first real attempt. Put the tool to work on the actual workflow — not a demonstration, not a hypothetical, the real thing. Have the person who owns that workflow use it on live work, with you or a trusted colleague close enough to watch. Expect it to be rough. The first attempt is always rough. That is not failure; that is week two doing its job.

Week three is for refinement. Now you adjust. You learn how to ask the tool for what you actually want. You discover which parts of the workflow it handles beautifully and which parts still need a human hand. The gap between a disappointing AI result and an excellent one is, more often than not, just a matter of learning to direct it well — and week three is where that learning happens.

Week four is for honest measurement. At the end of the month, you answer one question without flinching: how much time did this actually save, and was the quality as good or better? If the answer is genuinely yes, you have your proof, and you are ready for the next phase. If the answer is no, you have spent one month and a small amount of money to learn something true — which is exactly why the framework asked you to start small.

Notice what this month does not contain. No capital expenditure. No new hardware. No reorganization. No company-wide rollout. The Foundation phase is deliberately the cheapest, smallest, lowest-risk thing you will do in the entire ninety days — because its purpose is not transformation. Its purpose is proof.

What This Looks Like in a Real Business

Let me make this concrete, because the principle is easier to trust once you have watched it work.

Picture a small business — a couple dozen employees, the kind of operation most readers of this book will recognize. One of its most capable people, someone whose judgment the business genuinely depends on, was spending around twelve hours a week on a single recurring documentation task. Twelve hours. A day and a half, every week, of a skilled person's time, spent on routine paperwork that had to be done and that no one had ever stopped to question.

In the first thirty days — nothing more than the Foundation phase, exactly as described in this chapter — that one workflow went from roughly twelve hours a week to about three. Call it nine hours a week recovered. Over a month, that is something like thirty-six hours — close to a full working week of a skilled person's time, given back, from one workflow, in one month, at almost no cost.

> *One workflow. Thirty days. Nine hours a week given back to a person the business could not afford to keep buried in paperwork.*

Sit with what that nine hours represents. It is not just a number on an efficiency report. It is nine hours a week that a valued person now spends on work that genuinely needs them — the work that requires their judgment, their relationships, their experience. The Foundation phase did not replace that person. It rescued them from the part of the job that was wasting them. And it did so in a way small enough, cheap enough, and visible enough that everyone in the business could see plainly that it had worked.

That is the whole purpose of the first thirty days. Not a transformed company. One proven workflow, one freed-up person, and one piece of evidence solid enough to build the rest of the ninety days upon. Which is exactly where the next chapter goes — from a single proven workflow to AI woven into the way your business already runs.

CHAPTER FIVE

Days 31–60
Integration

From One Win to a Working System

You have a workflow humming. Your team has seen it work. One of your people has hours back in their week that they did not have thirty days ago, and the proof is no longer a theory – it is sitting in your own operation. That is exactly where the first month was supposed to leave you, and it is a real accomplishment. Do not rush past it.

But a single proven workflow, left by itself, is fragile. It is one good habit in a business still mostly run the old way. If you stop here, one of two things tends to happen. Either that single win slowly fades back into the routine as attention drifts, or it stays an island – the one place AI lives, disconnected from everything around it. The second thirty days exist to keep either from happening. This is the Integration phase, and its job is to take the thing you proved and weave it into the way your business actually runs.

What "Integration" Actually Means

The word integration can sound like an IT project — servers, contractors, a six-figure invoice. Set that picture aside. For a small or mid-market business, integration in the sense this chapter means is simpler and far less expensive than that. It means taking the AI capability you proved in month one and connecting it to the systems, the tools, and the work that surround it — so that it stops being a separate thing your team has to remember to use and becomes part of the natural flow of the day.

Think about what your first workflow looked like at the end of the Foundation phase. Almost certainly it still had seams. A person finished one step, then carried the result by hand to the next tool. They copied something out of one place and pasted it into another. They remembered, each time, to open the AI tool and do the thing. It worked — but it worked because a person was holding the pieces together. Integration is the work of removing those seams, one at a time, so the workflow holds together on its own.

> *Foundation proved AI can work in your business. Integration makes it part of how the business works — so it no longer depends on someone remembering to use it.*

Notice the difference in ambition between the two phases. Month one asked a single question: can this work here at all? Month two asks a different one: can this become durable? A capability that depends on a person's memory and goodwill every single day is not yet durable. A capability built into the workflow – so that using it is simply how the work is now done – is. That durability is what the second thirty days are buying you.

— ❖ —

The Three Moves of the Integration Month

Integration is not a single act. It is three related moves, and most businesses will work through them roughly in order over the second month. You do not need all three. You need as many as your one workflow genuinely calls for – and the discipline of month one still applies. You are deepening one workflow, not opening five new ones.

Move One: Close the Handoffs

The first move is to find the manual handoffs in your proven workflow and close them. A handoff is any point where a human being is acting purely as a courier – carrying information from one step to the next, copying from one system into another, moving a result along not because judgment is required but simply because the two pieces are not connected.

These handoffs are easy to overlook because they feel like work. They are not. They are friction dressed up as work. In the Foundation phase you tolerated them, rightly, because month one was about proving the concept, not polishing it. Now you go back and ask, at each seam: does a person need to be here because a decision is being made, or only because nothing is connected? Where the honest answer is the second one, that is a handoff to close — by linking the tools directly, or by letting the AI carry the result to the next step itself.

Every closed handoff does two things. It returns a little more time, and it removes a little more fragility — because a handoff that depends on a person remembering to do it is a handoff that will, eventually, be forgotten.

Move Two: Shift From Reactive to Proactive

The second move is the one that changes how Integration feels rather than just how efficient it is. In the Foundation phase, your AI tool did what you asked, when you asked it. It was reactive — a capable assistant waiting for instruction. The second move is to let some of that capability become proactive: to have AI watch for things and tell you, rather than waiting to be asked.

Most businesses are full of information that only gets looked at once something has already gone wrong. The numbers that reveal a problem are usually sitting in a system the whole time — they simply are not being read until the problem surfaces on its own. A customer account drifts toward trouble. A figure moves outside its normal range. A pattern in the work shifts in a way that matters. In the old way of running things, all of that is noticed late, by accident, when it has become urgent.

AI is genuinely good at this kind of watching, because the underlying skill is pattern recognition. You let it learn what normal looks like in some part of your business, and then it flags what does not fit — early, while there is still room to act. Instead of discovering a problem on Friday when it has become a crisis, your team gets a quiet note on Tuesday: this is trending in a direction worth a look. That is the shift from reactive to proactive, and it is one of the most valuable things AI does for a business that has been running on triage — handling whatever is on fire today and pushing everything else to a next quarter that never quite arrives.

> *Most businesses notice their problems late, by accident, once they have become urgent. AI is good at noticing them early — while there is still room to act.*

Move Three: Build the Checkpoints

The third move is the one that protects everything else, and it is the move an owner of conscience will care about most. As you close handoffs and let AI watch and act, a workflow that once passed through several pairs of human hands becomes something closer to a pipeline. That is the point. But a pipeline with no human checkpoints is not efficiency. It is a risk you have not yet noticed.

So as you integrate, you deliberately decide where the human checkpoints belong — and you keep them there on purpose. The principle is one the chapter on objections will take up in full, and it does not change: AI handles the repetition, and a person makes the decisions that require judgment. The goal of the Integration month is a workflow with fewer manual touchpoints than it had sixty days ago — fewer, not zero. The touchpoints you keep are not leftover inefficiency. They are the places where a human being looks at the work and takes responsibility for it before it goes any further.

Decide those checkpoints consciously rather than letting them fall wherever automation happens to stop. Put them where the consequences are real — before anything reaches a customer, before anything becomes a commitment, before anything carries your name out into the world. A well-integrated workflow is not one with no people in it. It is one where the people are positioned exactly where their judgment matters and freed from everywhere it does not.

The Goal for Day Sixty

Here is the simple test for whether the Integration month succeeded. At the end of it, you should be able to take your first workflow — the one you proved in the Foundation phase — and describe it out loud, start to finish, and have it come out shorter than it would have sixty days ago.

Fewer steps. Fewer seams. Fewer moments where a person is acting as a courier instead of a decision-maker. Perhaps a new line that did not exist before — the part where AI now watches something and raises its hand early. And running through it, at the points you chose deliberately, the human checkpoints where judgment still lives. If you can describe that workflow and hear that it has genuinely gotten leaner and more reliable, the second month did its work.

And notice what you now have that you did not have on day thirty-one. After the Foundation phase you had a win. After the Integration phase you have something sturdier — a working system. A piece of your business that runs better, more reliably, with less friction, and that will keep doing so without anyone having to remember to make it happen. A win can fade. A system tends to hold.

> *After month one you have a win. After month two you have a system — and a win can fade, but a system tends to hold.*

The Temptation to Resist

There is a specific temptation that arrives in the second month, and it is worth naming so you can see it coming.

By day forty-five or so, the first workflow is running well, the team believes, and the early skepticism has turned into something closer to enthusiasm. That enthusiasm is good — it is what you worked for. But it carries a familiar danger. The same impulse that once said buy six tools now says: this is going well, let us start three more workflows at once. The success of month one can talk you straight back into the breadth that kills AI efforts.

Hold the line. The Integration month is for deepening one workflow into a dependable system, not for spreading attention across new ones. There will be a time to widen — that is much of what the third phase is about, and there is real reward waiting in it. But you widen from a position of strength, with one fully integrated system already behind you, not from a scatter of half-built efforts. The discipline that made the first month succeed is the same discipline that makes the second one count. Faithful with one. Then the next.

Hold that line through the rest of the month, and you arrive at day sixty with something most businesses experimenting with AI never reach — not a pile of trials, but one genuine, integrated, durable system. That is the platform the final phase is built on. Because the first sixty days were about saving time and making the business sturdier. The last thirty are about something different. They are about taking the capacity you have created and turning it into growth – and that is where the next chapter goes.

CHAPTER SIX

Days 61–90
Expansion

From Saving Time to Making Money

For sixty days you have been playing defense. That is not a criticism — it is exactly what the first two phases were supposed to be. You found a costly workflow and proved AI could improve it. Then you integrated that win into a durable system. Both of those moves were about protecting something: protecting time, protecting your people from work that wasted them, protecting the business from friction and fragility. Defense well played.

The last thirty days are different. The Expansion phase is where you stop playing defense and start playing offense. You have cut waste and you have freed up capacity. Now the question is no longer how do I save time. It is the better question, the one every owner actually wants to be asking: what do I do with the capacity I have created?

> *The first sixty days were defense — protecting time and steadying the business. The last thirty are offense. The question changes from how do I save time to what do I build with it.*

This is the phase that completes the loop. Save time, reinvest time, make money. If the framework stopped at day sixty, you would have a leaner, sturdier business — genuinely worth having, but only half the point. The full point is what this chapter is about: turning recovered capacity into growth.

— ❖ —

Why You Could Not Have Started Here

It is worth pausing to see why Expansion comes third and not first. Almost every owner, reading the words make money, will feel the pull to skip straight to this phase. Resist that pull, and understand why it exists.

You cannot chase growth while your team is buried in operational overhead. An owner whose people are still consumed by repetitive, friction-heavy work has no capacity to point at anything new — every hand is already full. That is the trap the framework was designed around. The reason the first sixty days come first is not ceremony. It is that those sixty days are what buy you the bandwidth to innovate at all. Defense is not the opposite of growth. Defense is the thing that makes growth possible.

So when you arrive at the Expansion phase, you arrive carrying something an owner two months ago did not have: real, recovered capacity, and a team that has learned, through one proven system, that AI is a tool they can trust rather than a slogan they have to tolerate. That combination – capacity plus credibility – is the platform. You could not have stood on it on day one. You can now.

— ❖ —

Two Directions for the Capacity You Created

Expansion can move in two directions, and most businesses will pursue some of both. They are different in kind, and it helps to name them clearly.

Direction One: Widen — Apply the Framework Again

The first direction is the natural one. You have run the framework once, on a single workflow, all the way through Foundation and Integration. You now know how to do it. So you do it again – you take the next workflow on the list, the second-most-costly time thief your operators named back in the Foundation phase, and you put it through the same three steps.

This is where the discipline of the earlier chapters finally pays a dividend. Through sixty days you held the line at one workflow, and more than once that restraint felt almost too cautious. The Expansion phase is the release of that restraint – but a controlled release. You widen now because you have earned the right to: one system already proven, a team that believes, and a method you have run start to finish. You are not scattering attention across six untested fronts. You are repeating a process that worked. That is an entirely different thing, and it is safe in a way that the early scatter never was.

Widening also gets faster each time. The first workflow taught you and your team how the framework feels — how to choose, how to direct an AI tool well, where checkpoints belong. The second workflow inherits all of that learning. What took a careful ninety days the first time can take considerably less the second, and less again the third, until applying the framework becomes simply how your business adopts any new capability. That is the quiet, compounding reward of having started small.

> *You held at one workflow for sixty days. Expansion is the release of that restraint — but a controlled one. You widen because you have earned the right to, not because you got excited.*

Direction Two: Reach — Pursue Something New

The second direction is more ambitious. Widening applies AI to work your business already does. Reaching uses your recovered capacity to do something your business could not do before at all.

Every business has a list of things it would pursue if only it had the room — a new service line, a market it never had time to develop, a kind of customer it could not properly serve, a quality of work it always wanted to reach for but never had the hours to attempt. That list did not get done, in most cases, for one honest reason: there was no capacity. The people were fully consumed by the work already on the table.

The Expansion phase is where you take something off that list. Not the whole list — the discipline still holds — but one item. You use the time AI has given back to put one genuine growth initiative into motion. And here the goal of the framework is deliberately specific and deliberately modest.

The Goal for Day Ninety

Here is the target, and it is worth stating exactly, because owners tend to either aim too low or too high.

By day ninety, you should have at least one new initiative — one widened workflow well underway, or one genuine growth project in pilot — that was made possible by the capacity you created in the first sixty days. One. In pilot. Not finished, not fully scaled, not proven beyond all doubt. Underway.

Be careful with both halves of that target. The owner who aims too low treats day ninety as a finish line, banks the efficiency gains, and quietly lets the offense never begin — ending with a leaner business and an unrealized opportunity. The owner who aims too high decides the Expansion phase means a complete transformation in thirty days, overreaches, and recreates the exact too-much-too-fast failure the whole framework was built to prevent. The honest target sits between them: one initiative, genuinely begun, by day ninety. That is enough. That is, in fact, a great deal.

> *By day ninety: one new initiative, made possible by recovered capacity, genuinely underway. Not finished. Begun. That is the target — and it is enough.*

Because one initiative begun is not really one initiative. It is proof of a permanent new capability. It demonstrates — to you, and to your team, and to anyone watching the business — that this organization can now create capacity and reinvest it into growth. That is not a one-time event. That is a flywheel. Save time, reinvest time, make money, and the making of money frees still more capacity to reinvest again. The ninetieth day is not the end of the loop. It is the first full turn of a wheel meant to keep turning.

Make It Permanent: Capture What You Learned

There is one more task that belongs in the Expansion phase, and it is the one most likely to be skipped — because it produces nothing visible and feels like overhead. Skip it anyway at your cost. This is the work of systematizing: capturing what your business has learned so that the knowledge belongs to the business and not just to the few people who happened to be in the room.

Over ninety days, your team has accumulated something valuable and almost entirely undocumented. They have learned which tools suit which kinds of work. They have learned how to direct AI well — how to ask for what they actually want. They have learned where the human checkpoints belong and why. Right now, all of that lives informally, in the heads of the people who did it. That is fragile. People change roles. People leave. Memory fades. Knowledge that is never written down is knowledge the business can lose.

So before the ninety days close, write it down. It need not be elaborate — a plain, practical record of how your business does this: the workflows you have changed, the tools you settled on, the way you choose what to tackle next, the checkpoints you insist on. Call it a playbook. Its purpose is to turn ninety days of hard-won, personal experience into a permanent, transferable asset — something a new employee can be handed, something the business owns outright.

There is a stewardship instinct underneath this, and it is the same one that runs through the whole framework. You did not build your business to be a thing that works only while you personally hold all the pieces in your head. You built it, at least in part, to last — to be handed forward, to outlive any single person's memory, perhaps to be passed to the next generation. Writing down what you have learned is simply that intention applied to this new capability. It is how a ninety-day effort becomes part of the business's permanent character rather than a season it once went through.

Ninety Days Later

Stand at day ninety and look back at what the three phases actually built.

In the Foundation phase you proved, in your own business, that AI could take a real and costly workflow and make it dramatically better. In the Integration phase you turned that single win into a durable system — something that holds on its own, woven into how the business runs. And in the Expansion phase you took the capacity all of that created and pointed it, deliberately, at growth: one workflow widened or one new thing genuinely begun, and a playbook that makes the whole capability permanent.

You did not transform the company in a weekend. You did something steadier and more lasting. You moved one careful step at a time, let each result earn the next decision, and at the end of ninety days you are standing in a business that is leaner, sturdier, more capable, and — most importantly — one that now knows how to keep getting better. The framework was never really about ninety days. It was about teaching a business a way of adopting what is new without losing what made it worth building. Ninety days was simply long enough to prove it could.

The framework is now whole in your hands — three phases walked, with the AI DNA Blueprint from Chapter Three supplying the context every one of them runs on. What remains is the question every careful owner should still be asking: does any of this actually work in real businesses like mine? That is a fair question, and it deserves more than my word. The next chapter answers it with proof.

CHAPTER SEVEN

Proof It Works

The Framework in Real Businesses

Everything in this book so far has been a claim. A reasonable claim, I hope, and an honestly argued one — but a claim nonetheless. The framework works. The Blueprint matters. The ninety days deliver. You are right to want more than my word for it.

So this chapter offers something sturdier than argument. It offers proof — four accounts of the ideas in this book at work in real businesses, doing real things. I have changed the names and softened the exact figures, because the businesses and the people in them deserve their privacy and a published book is no place for another company's precise numbers. But the work is real, the results are real, and nothing here is invented or embellished. What follows happened.

I have deliberately chosen four stories that are unlike one another. One is a complex, high-stakes piece of business development. One is quiet, repetitive operational work. One is the routine creative output every business has to produce. And one is not a finished result at all, but a window into how the framework bends to fit an industry entirely different from the others. Taken together, they are meant to answer a single question: not does AI work somewhere, but could it work here, in a business like mine.

— ❖ —

CASE ONE · HIGH-STAKES BUSINESS DEVELOPMENT

The Bid That Would Not Have Been Built

The first story belongs to a vegetation-management company — a small, regional operation that clears and maintains land, the kind of unglamorous, essential business that exists in every part of the country. Our firm supported them on a piece of work that was, frankly, larger than anything they had taken on before.

A major renewable-energy developer had built a large solar facility nearby, and that facility needed its land managed — a substantial, recurring contract, the kind that can change the trajectory of a small company. But to win it, the vegetation company had to produce a formal bid: a complete, professional proposal package, properly scoped, properly priced, properly presented, the sort of document a large corporate client expects before it will take a smaller vendor seriously.

Here is the honest difficulty. A small operation like this one is superb at the actual work — the equipment, the crews, the land. What it does not have is a business-development department.

Producing a polished, comprehensive bid document is a different discipline entirely, and the realistic alternatives were grim: spend many days the owner did not have wrestling it together, or hand it to an outside firm at a cost that ate into the job before it was even won. For a great many small businesses, that is precisely the moment a real opportunity quietly slips away — not for lack of capability to do the work, but for lack of capacity to compete for it.

> *The opportunity was not lost for lack of skill at the work. It was nearly lost for lack of capacity to compete for it. That gap is where AI changed the outcome.*

This is where AI changed the outcome. Working from the company's genuine knowledge — its real capabilities, its actual pricing, its understanding of the site and the region — AI was used to build the full bid package: the scoping, the structure, the professional language, the complete proposal a corporate client would respect. Not generated from nothing. Built from the company's own substance, organized and expressed by AI into a form the company could not have produced alone at any speed it could afford.

The result was a bid the company could stand behind — and it became the basis of a real negotiation, one that moved from a single one-time job toward a recurring annual relationship worth considerably more. Notice carefully what AI did and did not do here, because it is the lesson of the whole chapter. AI did not win the contract; the company's actual competence and the owner's judgment did that. What AI did was let a small business compete at a level its size would normally have shut it out of. It closed the gap between being good at the work and being able to win the work. That is not a small thing. For a company that size, it was the difference between a door open and a door closed.

— ❖ —

CASE TWO · REPETITIVE OPERATIONAL WORK

The Reconciliation Nobody Wanted to Do

The second story is the opposite of the first in every way, and that is exactly why it belongs here. The first case was dramatic — a big opportunity, a high-stakes bid. This one is the quiet, grinding, invisible work that no business can avoid and no one enjoys.

This is one of our own companies, an energy brokerage. Like many businesses, it lives across several different record systems that do not naturally agree with one another: one system for customer relationships, separate reports from suppliers, separate files again for commission payments. And every so often, those systems have to be reconciled — matched against each other, line by line, to answer plain but essential questions. Which customers are active but were never paid. Which accounts have quietly closed. Where the records have drifted apart.

Done by hand, this is exactly the kind of task this book has been describing from the first chapter: repetitive, painstaking, error-prone, and a genuine thief of skilled hours. It is mechanical work — careful matching of identifier against identifier across thousands of rows — and mechanical work is precisely what AI is built to absorb. So we put AI to that reconciliation, comparing the systems against one another, sorting the results into clear and usable categories.

But this case carries the most important lesson in the chapter, and it is one this book has been honest about all along. The first pass was *wrong*. AI made a real error — it misread a particular kind of identifier number and produced a result that was confidently incorrect. The mistake was caught because a human being who genuinely knew the business looked at the output and saw that the numbers did not feel right. The error was identified, the cause was corrected, and the second pass was reliable.

> *The first pass was wrong. A person who knew the business caught it. That is not the system failing — that is the system working exactly as designed.*

Do not skip past that. It is not a flaw in the story; it is the heart of it. This is precisely the human checkpoint that Chapter Five insisted on — and that Chapter Eight will return to — the principle that AI handles the repetition while a person makes and verifies the judgments that matter. AI did the immense, tedious labor of matching thousands of records, work that would have consumed a skilled person for many hours. A human being supplied the experienced eye that knew when something was off. Neither could have done the job well alone. Together, a draining multi-hour task became a fast and dependable one — and the lesson is not that AI is perfect. It is that AI plus an informed human is powerful, and that the checkpoint is not bureaucracy. It is the design.

CASE THREE · ROUTINE CREATIVE OUTPUT

A Week of Writing in an Afternoon

The third story is about the kind of work that sits between the other two — not a rare high-stakes event, not invisible back-office labor, but the steady creative output that a business has to keep producing simply to function: the writing.

Every business has a version of this. The emails that have to go out. The proposals that have to be written. The marketing material, the customer communications, the recurring content that keeps a business present and professional in front of the people it serves. It is real work, it requires genuine thought and care, and it has a way of expanding to consume whatever time is available — which means it is often the work that gets done late, or rushed, or quietly not done at all.

The specific example was a complete welcome sequence — the series of messages a business sends a new customer to begin the relationship well. Done properly, a sequence like this is a meaningful piece of writing: it has to carry the voice of the business, set expectations, build trust, sound like a real person who genuinely cares that this customer chose them. Written from a blank page, it is the better part of a week of focused effort. It is also, for most owners, the kind of project that sits on the list for months precisely because a free week never appears.

With AI — and, importantly, with AI that had been given real knowledge of the business, its voice, and its customers, in the manner of the Blueprint described in Chapter Three — the substantive draft of that sequence came together in a single working session. Not a week. An afternoon.

> *With AI given real knowledge of the business, a sequence that would take the better part of a week came together in a single afternoon — still reviewed, still edited, still the owner's.*

Two cautions keep this story honest, and they matter. First, the speed was not magic; it was context. Generic AI, with no understanding of the business, would have produced generic writing and saved far less. The afternoon was possible because the AI had been taught the business first — which is the entire argument of the Blueprint chapter, shown here in practice. Second, the afternoon produced a strong draft, not a finished, sent product. A human being still read it, still edited it, still applied judgment and made it truly right before it ever reached a customer. The checkpoint held here exactly as it did in the reconciliation story. What changed was not the need for human craft. What changed was that the human began from a strong draft rather than a blank page — and the week that project would have cost became available for something else.

CASE FOUR · THE FRAMEWORK IN A DIFFERENT WORLD

When the Business Is a Whole City

The first three stories were results — things that happened and finished. The fourth is different, and I want to be plain about that from the start. This is not a completed engagement with a measured outcome. It is something else, and something useful in its own way: a window into how the framework in this book bends to fit an organization utterly unlike the small and mid-market businesses the rest of these pages address.

The organization in question is a destination marketing organization — the kind of public-facing body responsible for promoting a city as a place to visit. Set beside a vegetation company or an energy brokerage, almost everything about it is different. It is publicly accountable. It does not have customers in the ordinary sense; it has visitors, and it has a large network of local tourism businesses whose success is, in effect, its own. It was facing a genuine surge of demand from major events on its horizon, with a small internal team that would have to meet that surge without simply tripling in size.

And yet — this is the point of including it — the framework of this book applied to that organization with almost no strain. The questions were the same questions. Where are the costly, repetitive workflows? For this organization, content production at volume, the translation of visitor materials for international audiences, the handling of a rising tide of inquiries. Where could AI create leverage? In exactly those places, and in one more that an ordinary business does not have: equipping that whole network of small partner businesses, many with no marketing staff at all, with simple AI tools of their own.

The Blueprint idea fit just as cleanly. An organization that public, that distinct, would need an AI genuinely taught *its* particular voice, mission, and audience before any AI tool could be trusted to speak for it — the brilliant-but-uninformed new hire from Chapter Three, in a setting a world away from where that idea was introduced. And the 90-day discipline held: start with one workflow, prove it, integrate it, expand. The same staircase, in an entirely different building.

> *A vegetation company and a whole city are not alike in any obvious way. But the questions the framework asks are the same in both — and that is the point of this book.*

That is the whole reason this fourth story belongs here, unfinished though it is. The first three cases prove the framework produces results. The fourth proves something arguably more important: that the framework is not narrow. It is not a trick that works only for a particular kind of company. It is a way of thinking about AI and a business — any business — and the reason it travels so well is that it was never really about the technology. It was always about the disciplined questions: what is costly, what is repetitive, where is the leverage, what is the smallest first step. Those questions are as true for a city as for a crew that clears land.

— ❖ —

What the Four Stories Have in Common

Stand back from the four accounts and the pattern in them is worth naming, because it is the pattern this entire book has been describing.

In none of these stories did AI replace a person. In every one of them, AI did the heavy, mechanical, time-consuming part of the work – the assembling, the matching, the drafting – and a human being supplied what only a human can: the judgment, the knowledge of the business, the experienced eye that catches the error, the final responsibility for the work. The bid still needed the owner's competence. The reconciliation still needed the eye that knew the numbers were wrong. The writing still needed the human hand that made it truly ready. The pattern never changed: AI carried the labor, and a person carried the judgment.

And in every case the gain was the same in kind, whatever its size. Capacity. A small company could compete for work its size should have shut it out of. A draining multi-hour task became a short one. A week of writing became an afternoon. Hours, days, whole weeks – returned to the businesses and the people in them, to be spent on the work that genuinely needed them. That is what this framework does. Not robots. Not replacement. Capacity, given back to people, so they can do their best and most human work.

These four are simply stories — evidence that the claims of this book are not theory. The framework that produced them is the same one laid out in these pages, available to you, and waiting for the first small step. The chapters that remain answer the objections you are already carrying — and then ask you to take it.

CHAPTER EIGHT

The Objections You're Already Thinking

Honest Answers to Four Hard Questions

By now you have seen the framework. You know what the first ninety days look like, you have met the businesses that walked through them, and somewhere in the back of your mind a list has been forming. Not a list of next steps. A list of reasons.

That is normal, and it is healthy. No owner who has built something real hands it over to a new idea without questioning it first. The questions you are carrying right now are the same four questions I hear in every Mastermind room, every consulting call, and every conversation that starts with a hopeful nod and ends with "but here is my concern."

So let us put them on the table. Not to dismiss them — they deserve better than that — but to answer them honestly, the way I would if you were sitting across from me with a cup of coffee and no salesperson in the room. There are four. Cost. Your team. The work itself. And the one nobody says out loud.

"This Is Going to Cost More Than I Can Justify."

This is usually the first objection, and it is the easiest to answer, because the numbers are not on the side of your fear.

The first thirty days — the Foundation phase, where you take a single workflow and put AI to work on it — can be done in most cases for under a thousand dollars. Often for far less. The tools that handle the kind of work most owners start with carry free tiers and low-cost plans that are more than sufficient to prove the concept. You are not buying an enterprise system. You are buying a month of one tool aimed at one bottleneck, and you are finding out whether it works before you spend anything more.

The Integration phase — days thirty-one through sixty — is where some real cost can enter, if you choose to connect AI more deeply into the systems you already run. Even then, the honest figure is low five figures at the outside, not six. And many owners never need to spend that, because the off-the-shelf tools have closed the gap.

But cost is the wrong number to stare at by itself. The right number is the one sitting next to it.

Run the arithmetic the way you would on any other investment. If putting AI on one workflow gives a key person back ten hours a week, that is not a soft benefit — it is real payroll, recovered. Ten hours a week is more than five hundred hours a year of skilled time that was being spent on work the business should never have been paying for by hand. Set the one-time cost of the tool against a year of that recovered time, and the question answers itself.

Most owners who implement honestly see the math turn positive inside sixty days. The cost objection, in other words, is real for about two months and then it inverts — and from that point forward the tool is paying you.

> *The cost objection is real for about two months. After that, the tool stops being an expense and becomes a line on the other side of the ledger.*

There is a deeper version of this objection, and it is worth naming. It is not really about a thousand dollars. It is about the fear of spending money on something you do not yet understand. That fear is legitimate — and the framework is designed precisely to retire it. You do not spend big and hope. You spend small, you watch one workflow, and you let the result earn the right to the next decision. Stewardship is not the avoidance of investment. It is the disciplined testing of it.

"My Team Will Never Go Along With This."

This is the objection I take most seriously, because the owners who raise it are usually right about the difficulty and wrong about the solution.

They are right that adoption is hard. They are wrong in how they picture the resistance. When an owner imagines team pushback, they tend to imagine the whole team, arms folded, as one wall. That is not how it works. A team is not a wall. It is a room full of individuals, and the truth is that you do not need the room. You need one person.

Start with the person who is most frustrated. I mean that literally. The team member who complains the most about repetitive, grinding, soul-flattening work is not your problem — that person is your opening. They have the most motivation to try something new, because they are the one feeling the pain every single day. Go to them, not with a mandate, but with a question: what is the one task that eats your week? Then put AI on exactly that.

Show that person a genuine win inside the first two weeks — hours given back, a dreaded task made small — and something changes that no memo could ever accomplish. They stop being a skeptic and become your internal evangelist. And here is the part owners underestimate: relief is contagious in a way that instruction never is. When the rest of the team watches a colleague get five hours of their week back, resistance quietly turns into something closer to envy. They stop asking whether they have to and start asking when it is their turn.

> *A team is not a wall. It is a room full of individuals — and you do not need the room. You need one person, and you start with the one who is hurting most.*

Two smaller fears usually travel alongside this one, and both deserve a straight answer.

The first is *"we are not technical enough."* Set that one down. The tools that matter here are easier to use than the spreadsheet software your team already runs without a second thought. If a person on your team can write an email, they can use these tools. The barrier was never technical skill. The barrier is simply starting.

The second is *"we do not have time to implement this."* I understand the feeling, but look closely at the logic, because it is backwards. The first use case takes a few hours to set up. You earn those hours back inside the first week. Implementation does not consume time. It is the thing that creates it. The owner who says there is no time to implement is describing exactly the condition that implementation is meant to cure.

Adoption, in the end, is a cultural challenge far more than a technical one. It does not spread through corporate announcements. It spreads through demonstrated relief — one freed-up person at a time. Lead it that way and the team comes with you, not because you required it, but because they watched it work.

"Can I Really Trust AI With the Work That Carries My Name?"

This objection sounds like a question about quality. It is really a question about identity, and that makes it the most important one in this chapter.

Can AI produce the work? In a great many cases, yes — it can draft, summarize, organize, and handle repetitive formats with a speed no human can match. But should it produce that work and send it out into the world untouched, with your name on it and no one having looked? Absolutely not. And the distinction between those two answers is the whole of the matter.

Think of AI as the most capable research assistant you have ever employed. It will produce excellent first drafts. It will compress hours of source material into something you can act on. It will take the repetitive, formatted work — the routine summaries, the standard documents, the predictable correspondence — and carry it for you. What it will not do, and must never do, is make the final judgment. The editorial decision, the one that says this is good enough to represent us, belongs to a human being. It belongs to you, or to someone you trust, every time.

> *AI is the research assistant, not the one whose name is on the door. It drafts. You decide. That line does not move.*

Your credibility is your brand. For most owners reading this, the business carries their name, their reputation, their word — and in many cases their faith and their family legacy are bound up in it too. That is not a thing to automate away in the name of efficiency. The goal of the framework was never to remove your judgment from the business. It was to remove the busywork that was burying your judgment, so that the discernment only you can offer finally has room to operate.

Used this way, AI does not dilute the work that carries your name. It does the opposite. By taking the rote labor off your best people, it frees them to spend their attention on the parts of the work that genuinely require a human — the craft, the relationships, the calls that need wisdom rather than speed. The tool handles what is mechanical. The people handle what is meaningful. That is not a compromise of your standards. It is a return to them.

— ❖ —

"Isn't This Just Going to Cost People Their Jobs?"

This is the objection nobody raises in front of the group. It surfaces afterward, quietly, one-to-one — because it is not a business question. It is a moral one, and the owners who carry it are usually the ones who care most about the people who work for them.

So let me answer it the way it deserves to be answered, plainly.

AI replaces tasks. It does not replace people. That sentence is easy to say and easy to wave away as a comforting slogan, so look at what actually stands behind it. Every business has work that is mechanical and work that is human. The mechanical work — the repetitive, the formatted, the predictable — is what AI is genuinely good at absorbing. The human work — the strategy, the creativity, the relationships, the judgment under uncertainty — is what AI cannot do and what your people are, in fact, uniquely there to do.

When you put AI on the mechanical work, you are not erasing a person. You are changing what their day is made of. The frustrated team member from earlier in this chapter — the one drowning in repetitive tasks — does not lose a job. That person gets a better one. The same role, but reoriented toward the work that is strategic, that is valued, that is worth a skilled human being's time. Across the businesses I have watched do this honestly, the pattern is consistent: the ones who implement AI well are not shrinking their teams. Many of them are hiring, because the business is now capable of more.

> *AI replaces tasks, not people. Put it on the mechanical work, and your team does not get smaller — it gets more capable, more valued, and freer to do work that is actually worth their time.*

But I want to go one layer deeper, because for many owners reading this the question is not really operational. It is a question of conscience. You feel a responsibility for the people who depend on the business you built, and you should. That responsibility is one of the better things about you.

Hold that responsibility up against the alternative honestly. The owner who refuses to adopt AI out of a sincere desire to protect jobs is making a real moral choice — but it may not be the protective one it feels like. A business that falls behind does not protect anyone. The competitor down the road who moved first does not share that hesitation, and the slow erosion of a business that would not adapt costs jobs too. It just costs them later, and more quietly, and with no chance to plan for it.

The genuinely responsible path is not avoidance. It is to adopt AI deliberately, to be honest with your team about what is changing and why, and to use the capacity you gain to build something more durable — a business that can still be there, still employing people, still serving the people it serves, ten years from now. Caring for your people and adopting AI are not opposing commitments. Done well, they are the same commitment.

The Real Question Underneath All Four

Look back at the four objections and you will notice something. Underneath the cost worry, the team worry, the quality worry, and the jobs worry, there is one shared root. Every one of them is, in its honest form, a fear of getting an important decision wrong.

That fear is not a weakness. It is the instinct of someone who takes the weight of the business seriously. But notice what the framework has been doing with it the whole time. It never asked you to make one large, irreversible bet. It asked you to take one workflow, give it thirty days, and watch. It asked you to start with one person, not the whole team. It asked you to keep your judgment on every piece of work that carries your name. It is built, from the first page, to let the results earn each decision before you make the next one.

You do not have to resolve every objection in this chapter before you begin. You only have to be honest that they are answerable — and then take the first small, testable step anyway. The next chapter is about exactly that step: what to do on Monday morning.

CHAPTER NINE

Your Monday Morning Move

What To Do With What You Now Know

A book about AI can do one of two things. It can leave you better informed — or it can leave you changed. The difference between those two outcomes is not decided by the book. It is decided in the next few days, by you.

Everything in these pages was built to be acted on. The 90-day framework, the AI DNA Blueprint, the honest answers to the objections you were already carrying — none of it was meant to be admired. It was meant to be used. And the gap where good intentions quietly die is always the same gap: the one between closing the book and doing the first thing. So this final chapter is short, and it is practical, and it has one job. To make sure you cross that gap.

I am not going to ask you to do something large. Large is the enemy — you have read a whole chapter on why. I am going to ask you to do something small enough that there is no honest excuse not to. Here is your first week.

Your First Week, in Three Steps

Not your first ninety days. Not your AI strategy. Your first week — the five working days that begin the moment you put this book down. It has three steps, and not one of them requires a budget, a meeting, or anyone's permission.

Step One. Have one conversation.

Before this week is out, sit down with one person on your team — one of the people whose hands are actually on the work — and ask them a single question: what is the one task that eats your week? Then be quiet and listen. Write down the answer. That answer is the seed of everything. You are not solving it yet. You are simply finding it, from the person who knows.

Step Two. Spend thirty minutes on one tool.

On Monday morning, before you open your email and the day takes you, spend thirty minutes — no more — looking into one AI tool suited to the task that person named. Not five tools. Not a comprehensive survey of the market. One. The point of this step is not the perfect choice. The point is a chosen one.

Step Three. Run one trial by Friday.

By the end of the week, have that tool in an actual trial run on that actual workflow. Not a committee meeting about whether to try it. Not a feasibility study. A trial — a real attempt, on real work, by the person who owns it. It will be rough. It is supposed to be. A rough trial completed beats a perfect plan untried, every time.

That is the whole of your first week. One conversation, thirty minutes, one trial. Read it again and notice how small it is — because the smallness is the entire point. There is no version of your schedule, no matter how full, that cannot contain those three things. Which means the only thing that can stop them is the decision not to begin.

> *One conversation. Thirty minutes. One trial run by Friday. Small enough that the only thing that can stop it is deciding not to begin.*

— ❖ —

How One Week Becomes Ninety Days

Here is what makes that small first week worth taking seriously: it is not small in what it leads to.

That first week is the opening of the Foundation phase. The trial you run by Friday is the beginning of the one workflow you will prove over thirty days. That first proven workflow becomes the thing you integrate into a durable system over the next thirty. And that system becomes the platform you expand from — widening to the next workflow, reaching for the growth you never had capacity for — across the final thirty. The first week becomes the first month. The first month becomes ninety days. And ninety days from the moment you put this book down, you are standing in a measurably better business than the one you run today.

None of that chain requires a heroic act anywhere along it. That is the quiet genius of working this way, and it is worth saying plainly one last time. You never have to make a single large, frightening, irreversible decision. You only ever have to take the next small step and let its result earn the one after it. The framework is not a leap. It is a staircase. And every staircase, no matter how high it climbs, is built the same way — one step, then the next.

> *The framework is not a leap. It is a staircase — and every staircase, however high it climbs, is built one step, then the next.*

The Ones Who Move First

There is a belief among small and mid-market owners that the AI advantage belongs to the big companies — that the businesses that win with this technology will be the ones with the largest budgets, the dedicated departments, the resources to do it at scale. I want to leave you with the truth, because that belief is wrong, and it is wrong in a way that matters.

The businesses that will thrive in the years ahead are not the ones with the biggest budgets. They are the ones that move first. Everything in this book has been designed around that fact. The framework costs little. It needs no department and no scale. It rewards exactly the things a smaller business already has in abundance and a larger one often lacks – the ability to decide quickly, to start on Monday without a quarter of approvals, to put a tool in a person's hands by Friday because the owner simply said so. Your size is not the disadvantage you may have feared it was. Used rightly, it is the edge.

Which means the question was never whether you have the resources. You do. The question is only whether you will move. The owner who reads this book, nods, agrees with all of it, and changes nothing will be overtaken – not by a competitor with more money, but by a competitor much like themselves who simply began. And the owner who takes the small first step, this week, with no more resources than they have right now, has already separated themselves from most of the field. Moving first is not a matter of means. It is a matter of will.

— ❖ —

You Do Not Have to Walk It Alone

One last word, and it is an honest one.

This book has given you the whole framework. The path is real, it is laid out plainly, and you can absolutely begin walking it on your own — starting this week, with the three steps above. I hope you do. Nothing in these pages was withheld, and nothing in them was meant to leave you dependent.

But I have spent a long time in rooms with business owners, and I know something about how this goes. The framework is simple to understand and genuinely harder to sustain alone. Momentum fades when no one is watching. Questions come up that a book cannot answer, because they are specific to your business in a way no book can anticipate. And the work goes faster, and holds better, in the company of other owners doing the same thing at the same time — people who can compare notes, who can tell you what worked in a business not unlike yours, who keep you accountable to the next step when the week gets loud.

That is the whole reason the work I do beyond this book exists. There is a community of owners walking this exact path together, and there is deeper, more personal help for those who want a guide rather than a guidebook. You will find the way to both at the back of this book. I mention it not as a sales pitch but as a simple, honest offer: if you would rather not walk it alone, you do not have to. The door is open, and the choice — like every choice in this book — is entirely yours.

> *You can walk this path alone, and I hope you begin this week if you do. But you do not have to walk it alone — and that choice, like every choice in this book, is yours.*

— ❖ —

Now Go and Begin

So here is where the book ends and your part begins.

You picked this book up, and you read it to the last page — which already tells me something true about you. You are not a person content to let the world change while you watch. You are the kind of owner who shows up, who finishes what they start, who takes the work of leading seriously. That instinct is the only qualification this framework ever required. You already have it. You always did.

The technology will keep moving whether any of us is ready or not — that much is not in your hands. But what you build with it, in your business, with your people, is very much in your hands. You have been given a clear path and a small, honest first step. The capacity you create with it, the people you free to do better work, the more durable business you hand forward to whoever comes after you — all of that is still unwritten, and you are the one holding the pen.

Have the conversation this week. Spend the thirty minutes on Monday. Run the trial by Friday. Then take the next step, and the next. Ninety days from now you will be glad you began — and the only way to get there is the same as it has always been.

Now go and begin.

AFTERWORD

If You'd Rather Not Walk It Alone

An invitation — and where to go next

You have the whole framework now. Everything in this book was given plainly, and you can begin walking the path on your own — this week, with the three steps in the final chapter. I hope you do.

But as the close of this book said honestly: the framework is simpler to understand than it is to sustain alone. Momentum fades when no one is watching. Questions come up that are specific to your business in ways no book can answer. And the work goes faster, and holds better, in the company of other owners doing the same thing at the same time.

That is why the work beyond this book exists — not as an upsell, but as a genuine offer for those who would rather have a guide than a guidebook. There are two ways to go further:

The AI Business Briefing

A regular, no-cost briefing for owners putting these ideas to work — the place to start if you simply want to stay close to the conversation. Visit MSGResources.com/briefing

The Mastermind & Private Advisory

For owners who want structured, accountable help walking the 90-day path — a working group of peers, and, for those who want it, deeper one-to-one guidance. MSGPR.com/ai

Whichever you choose — including choosing neither and walking the path alone — the decision is yours, exactly as every decision in this book has been. The door is simply open.

Talk It Through Right Now

If you finished this book with a question specific to your business, you do not have to wait for an answer. I have trained an AI version of myself on everything I teach — ask it anything, any hour, at MSGResources.com/FAQ. It is, fittingly, the AI DNA Blueprint at work.

About the Author

Lee Allen Miller is a faith-driven entrepreneur based in East Texas and the founder and CEO of MSGPR Ltd Co. Over more than forty years in business, he has built and led ventures across media production, marketing, energy, broadcast technology, and publishing. He is the bestselling author of *Entrepreneurship God's Way* and *The Faith Media Revolution*, and he leads an AI education practice helping small and mid-market business owners adopt artificial intelligence with discipline and wisdom.

He writes and works from Lufkin, Texas. Connect with him at MSGPR.com.

❖

www.ingramcontent.com/pod-product-compliance
Lightning Source LLC
LaVergne TN
LVHW022013200926
848481LV00007B/2372
9798992107432